D1156673

THE DANGERS OF NOISE

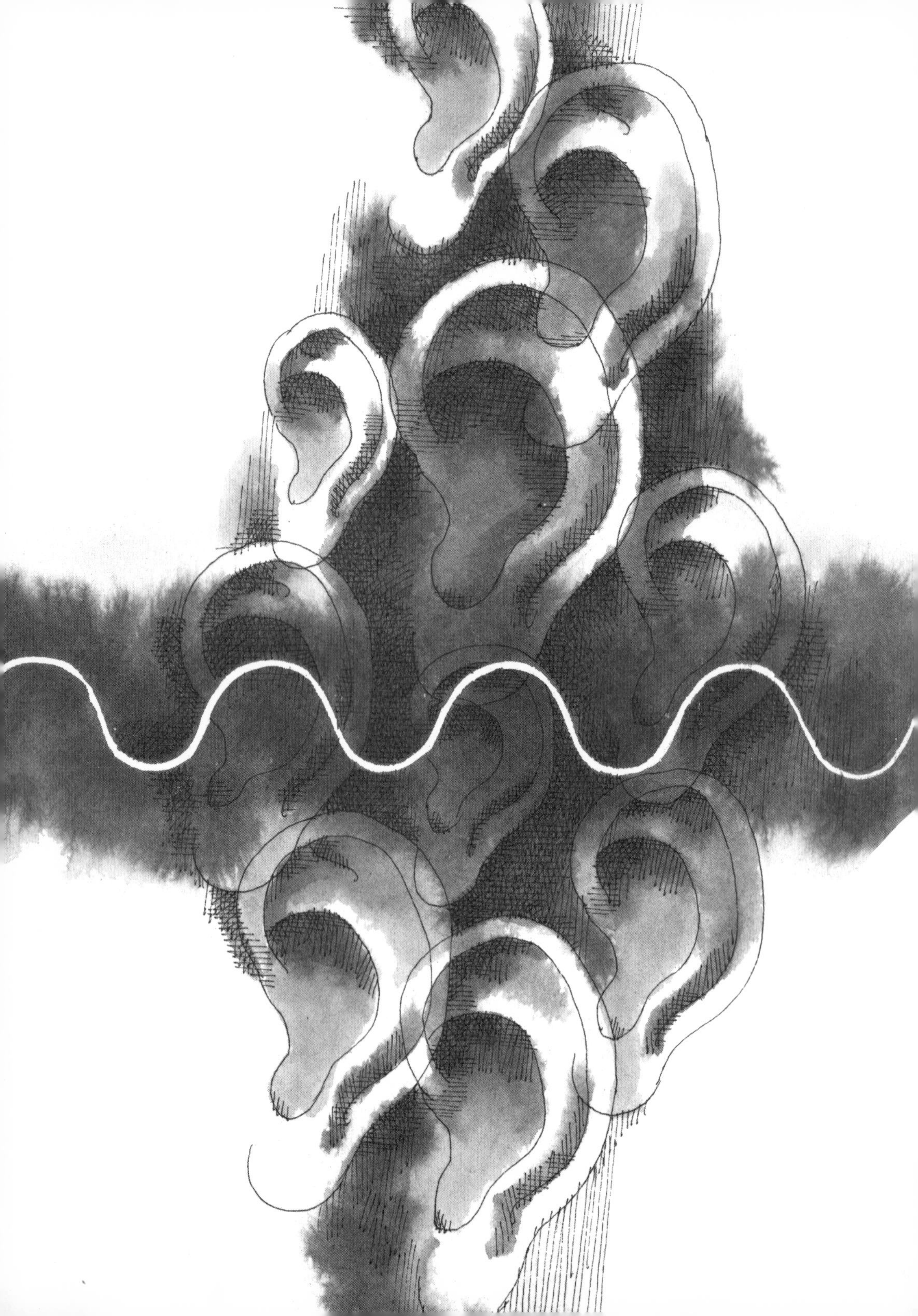

THE DANGERS OF NOISE

by LUCY KAVALER

Illustrated by Richard Cuffari

Thomas Y. Crowell New York

 Published simultaneously in Canada by
Fitzhenry & Whiteside Limited, Toronto.
For information address Thomas Y. Crowell,
10 East 53 Street, New York, N.Y. 10022.
Manufactured in the United States of America

Library of Congress Cataloging in Publication Data

Kavaler, Lucy. The dangers of noise.
SUMMARY: Discusses the seriously detrimental effects
of noise on people and the environment.
1. Noise pollution—Juvenile literature. 2. Noise
—Physiological effect—Juvenile literature. [1. Noise
pollution. 2. Noise—Physiological effect] I. Cuf-
fari, Richard. II. Title.
TD892.K32 363.6 77-26588
ISBN 0-690-03905-0
ISBN 0-690-03906-9 lib. bdg.

To Roger and Andrea

OTHER BOOKS BY LUCY KAVALER

Noise: The New Menace
Freezing Point: Cold As a Matter of Life and Death
Mushrooms, Molds, and Miracles
The Astors: A Family Chronicle of Pomp and Power
The Private World of High Society

FOR YOUNG READERS

Life Battles Cold
Cold Against Disease
Dangerous Air
The Wonders of Fungi
The Artificial World Around Us
The Wonders of Algae
The Astors: An American Legend

Contents

one	Noise, the Invader	1
two	How Loud Is Too Loud?	8
three	The Strange Ways of Sound Waves	16
four	Attack on the Ear	24
five	What Noise Is Doing to You	32
six	Animals Are Victims, Too	39
seven	The Laws of Silence	47
eight	The Noisiest Way to Travel	55
nine	Search for a Quiet Machine	62
ten	You Can Say No to Noise	69
	Index	79

THE DANGERS OF NOISE

chapter one
NOISE, THE INVADER

The Knights Hospitalers of St. John of Jerusalem came from their homeland to the beautiful island of Rhodes in the Aegean Sea in the year 1309. True to their name, one of their very first acts was to build a hospital. They were horrified by the noise in the bustling town of Rhodes and thought it would bother the sick people and slow their recovery. The hospital was, therefore, built around a courtyard left open to the sky. The side facing the street was a solid wall of thick stones with a few tiny windows set high up. Light and air came through the windows on the court. Little sound from the streets reached the patients in their beds.

Noise is not a new problem; it has been troublesome since people first gathered together in cities. The Knights Hospitalers devoted themselves to the care of the sick, but those who are healthy are also disturbed by noise. Sybaris was a Greek colony founded during ancient times in southern Italy. It was a rich colony, and the

Sybarites became known for their love of comfort and the beautiful things of life. Early in the eighth century B.C. the Sybarite government ordered manufacturers to move to the outskirts of the city. Metalworking shops were to be located the farthest away from homes so that the pounding would not interrupt conversation and drown out song. Those who wished to sleep late in their soft beds were encouraged to do so. No one living within city limits was allowed to keep a rooster to break the early morning quiet with its loud cry.

Today hardly anyone is as free from noise as the ancient Sybarites. Most of us live in cities or suburbs where noise is so constant that we wake up at night if it stops, startled by the sudden unfamiliar silence. The sound of jet planes, trucks, and drills follows us on weekends and on vacation. Climb to the top of a mountain, and the sound of the traffic on the roads below will rise to greet you.

What can all the noise be doing to people? It not only annoys, but also dulls the sense of hearing. The ears are battered by sound day after day, year after year, and in time the hearing mechanism grows weary. Sounds and words become blurred. It is hard to follow conversations, and the television set and stereo must be turned up high.

Noise also affects the rest of the body and the mind. It can make the world a more frightening place to live in. What happens when you hear a sudden loud sound? Your body reacts as if it expected to be attacked. Your ears hurt as they take the force of the blast. Is the sound a warning signal of danger? Do you have to run away or stand and fight? It is not surprising that people who are startled by noise too often become nervous and irritable, ready to quarrel.

Even animals react with fear. The cotton rat of the grasslands of Florida used to be both friendly and bold. That was before the coming of an airport, with airplanes taking off or flying in for a landing. Today the rats are timid, slow to creep out of a hiding place, and they dig their nests deep beneath the ground.

Over the centuries governments have tried to protect their citizens from noise. Sybaris was in time destroyed, but its ideas about noise lived on. The sound of chariots going down the streets of ancient Rome was so annoying to Julius Caesar and his senators that nighttime driving in the city was outlawed.

But it did not prove easy to control noise. The centuries passed, and the world became even less quiet. The ring of the anvil in the blacksmith's shop, the banging of the metalworker, the hammering of the shoemaker, bothered those living in the towns of the Middle Ages. Coming closer to the present, the number and size of factories grew, and new and noisier machines were invented. Today people, houses, factories, airports, and schools are crowded together.

Students in a school near the Los Angeles airport decided to do a noise project. They counted the number of times the teacher's voice was drowned out by airplanes passing overhead. Between eight thirty and noon, this happened sixty-five times. As a result, pupils could not learn their geography.

Certainly, before there were airplanes and trucks, before Sybaris was built, even before humankind evolved, the world was never quiet. Trees fell, animals croaked and roared, thunder crashed in prehistoric jungles and swamps though no human was there to hear. But was that noise or was it just sound? Can you tell the difference between the two?

A century ago British physicist John William Strutt decided that noise was "unmusical," and sound "musical." The creaking of a shoe, he said, was noise, while the playing of a piano sonata was sound.

But you could surely find many exceptions to that rule. Some

modern music makes use of city noises, of blares, creaks, and beeps. And what of rock? The rock band is making . . . noise . . . music? If that were a multiple-choice question on a test, most teen-agers would answer music, and most older people noise. And each of them would be right, from his or her own point of view.

You can see that although *noise* seems to be a very easy word, it is not really easy to get everyone to agree on what it is. A new way of defining it has been worked out. Noise is "unwanted sound." At first it may seem as if this were not a very good definition. Unwanted by whom? By you? By me? By your family? By an airplane pilot? The performer on an electric guitar? The owner of an automobile factory?

"There will always be some kinds of sound that people disagree about. The definition means unwanted by the majority," says an environmental-agency official. "The world would be much quieter if we just held down the sounds that annoy most people."

The roar of the motorcycle engine being charged is exciting to the rider, unpleasant to the crowd left standing on the sidewalk. The news is interesting to the person listening to it on the radio at home, annoying when it is turned up so high that all the neighbors must hear. The sound of a jet plane flying low over houses may be pleasant to the ears of the pilot or those of the aircraft maker, but it bothers many millions on the ground. The drill continuing to break up the street pavement until late at night may not disturb the owner of the construction company, but it makes life miserable for everyone living on the block.

Like the Sybarites and the Romans, modern Americans have passed a great many laws to make their cities quieter. Did you know that in New York City it is against the law for the driver of an ice-cream truck to play the jingle on the bell for longer than ten seconds? A tambourine may not be rattled on the streets of Illinois towns and cities. Police have been called out in Boulder, Colorado, to quiet the screeching of peacocks in the zoo. Wherever you live, there are laws saying that the sound of music and talk at a late-night party must not disturb other people. American factory workers must not be asked to spend more than fifteen minutes a day working at a very noisy machine. Drivers in Memphis, New York, and many other cities may not honk their horns unless they are warning someone else of danger.

Sometimes most people agree that a particular sound is unpleasant, and yet they also agree that they are lucky to hear it. The camper walking carelessly through the woods who suddenly hears a bear growl before it appears can jump away in time. A family is sleeping, and all at once a fire alarm goes off and wakes everyone up. You are about to step off the curb when you hear tires screech as a car skids around the corner, and so you stay where you are. Each of these unpleasant sounds is useful.

And so the noise-control laws use the principle that noise must mean sound that is not only unpleasant and too loud, but also unnecessary.

In order for the laws to work, each person must obey them and ask others to do so, too.

Some time ago the monkey house in a big city zoo was going to be enlarged. The animals were moved into the next building, and a team of construction workers arrived and prepared to begin blasting.

Just before the dynamite was set off, the zoo director rushed up. "You can't blast!" he shouted. "The apes will go wild, and I won't stand for it."

The contractor had to give up his plans to blast and find a quieter way to do the job.

Anyone can follow the example of the zoo keeper. A new apartment house was being built in a quiet section of a suburb. Evening came, and the crew was still at work. The families living in the houses nearby were getting desperate. Finally one young woman came out, went up to the foreman, and said, "Stop! This racket is driving us all crazy."

She had expected an argument. But the foreman only said cheerfully, "I'm glad you came. We were told to keep on working until someone complained."

chapter two
HOW LOUD IS TOO LOUD?

Where is the quietest place in the United States? Where is the noisiest? A scientist took equipment to measure sound in his pack and traveled over the countryside. And one day he came to the Grand Canyon and rested there on the north rim. There was no sound louder than that of his movements and breathing. Later he spent a day in a third-floor apartment in Los Angeles, right over a freeway and under the flight paths of the airplanes leaving the huge Los Angeles airport.

He hung his measuring device out of the Los Angeles window for twenty-four hours, checking the readings every hour, even during the night. When he averaged out the figures, the noise level came to nearly 89 decibels. At the Grand Canyon the reading was 12 decibels.

How much noisier is the third-floor apartment in Los Angeles than the Grand Canyon? The difference between the two is far greater than the difference between the numbers 89 and 12 might lead you to believe. The sound in the apartment over the freeway is roughly 999,999,990 times louder than that at the Canyon's rim.

The explanation can be found in the system of decibels, the units by which we measure sound. Decibels, abbreviated dB, are not like inches or ounces. You cannot add them together or subtract one from the other. Instead, they are figured according to a kind of mathematics known as common logarithms. To work out logarithms, you must divide and multiply by 10. First, you must find out how many times 10 goes into the decibel number given you for a certain sound. So you divide the number by 10 and when you get your answer, then you multiply. Following this system, you will find that 10 decibels of sound is 10 times as loud as 1 decibel (10 goes into 10 once, so 10 times 1 gives the answer); 20 decibels is 100 times as loud (two 10s in 20, so 10 times 10); and 40 decibels is 10,000 times as loud (10 times 10 times 10 times 10).

Because you need a calculator to get the exact logarithm of odd figures, you can round off the 89 decibels of sound at the Los Angeles apartment to the closest multiple of 10, which is 90. This gives you 9 tens (10 times 10 times 10 times 10 times 10 times 10 times 10 times 10 times 10) or 1 billion. Round off the 12 decibels of sound at the Grand Canyon to 10 and multiply (1 times 10). You subtract that 10 from a billion and get the 999,999,990 difference between the two sounds.

The decibel system begins at 0 to 1 decibels, the softest sounds the human ear can hear. You can get some idea of how soft that is from the fact that even the beat of your heart, the flow of blood through your veins and arteries, the movement of the digestive system, give out a few decibels. Your quiet breathing when you are sitting still makes 10 decibels of sound. Doctors are listening for the faint heart and lung sounds when they put a stethoscope to a patient's chest.

Because figuring decibels takes time until you have had some practice, it is a good idea to fix in your head the decibel counts for a few of the sounds you hear often. Then you can compare other noises with these.

In the wilderness the rustling leaves, the chirping birds, and scurrying animals make noise of 20 to 30 decibels. In a school library, students asking in soft voices for books, pulling them from shelves, and turning pages bring the sound level to 35 decibels. Two friends talking together in a lively way make about 50 decibels. Two students who are sitting at opposite ends of a table in the school cafeteria will be unable to hear each other very well; the dishes banging, the talk and shouts and movements, make a background sound level of at least 63 decibels. A girl with a light voice will barely be heard by someone sitting just two feet away. A boy's deep voice will carry for about four feet.

At night a snoring roommate can keep the other person awake. Loud snores send out 65 decibels of sound. The alarm clock that gets the sleeping student up in time for school does so with a shrill 80 decibels. An automobile traveling at a speed of 35 miles an hour is heard at 82 decibels by someone standing 50 feet away—and the sound is louder than that if the car is old and has a poor muffler.

A disco or a party where a rock band is playing with electrified instruments will give out a noise level equal to the greatest nature can produce—the 120 decibels of the thunderclap. At this 120-decibel level, most people feel uncomfortable, even if they think they are enjoying themselves completely. A sound of 130 decibels actually hurts. That is not the limit of noise people sometimes have to hear. The army reports that artillery fire at 500 feet gives 150 decibels of noise and a bomb explosion reaches a terrifying 190 decibels.

These decibels describe an amount of energy. This energy is created by the movement of the air. The best way to understand that is to think of sound as movement. Every time a dog barks, a guitar string sings, a bell rings, or a voice speaks, it sends out energy that makes the air around it move. Some voices and musical instruments make the air move violently, while others bring about more gentle movements.

These air movements are called sound waves, because they act very much like the waves in the ocean. If you ride an ocean wave, you are carried forward as the wave rises to a crest, then dropped down again, and then lifted forward. The waves of sound cannot be seen, but they move in much the same way. The voice or violin string pushes the air ahead for a certain distance until it reaches a peak, and then drops back, only to move forward again.

If you are in the path of the sound waves, they will reach your ears and enter through the hole in the outer shell. The part of the ear that is outside your head is the least important as far as hearing goes. Far more important are the middle ear and the inner ear, which are both inside your head.

Once they have entered the ear, the sound waves beat against the thin eardrum that is at the entrance to the middle ear. They make the eardrum shake and the middle ear begins to shake, too. As the sound waves approach the inner ear, two tiny muscles in the middle ear tighten so as to allow just the right amount of sound to get to the inner ear. They protect the delicate inner ear from being battered by too much sound. So when the sound of electrified guitar, drill, or cap pistol could do damage, the muscles close the entrance to the inner ear. But if the noise goes on too long, the muscles give up and relax, and the full force of the sound gets inside.

Some sound must reach the inner ear, if you are to hear it. The ear is formed like one of those sets of magic boxes in which you open one box only to find another inside, and then another inside that until finally you reach the treasure within. If you could get inside the magic box of the inner ear, you would find the cochlea (*cock-lee-ah*), shaped like a snail's shell. And within the cochlea lies the treasure—the organ of Corti. No larger than the tip of your fingernail, this organ is the most important part of your hearing.

If you look at the organ of Corti under a microscope, you will see something that may surprise you—rows and rows of hairs. These hairs are connected to tiny nerves that carry sound from the ear to the main nerve of hearing that runs to the brain, where the meaning of the sound is figured out. The brain also decides whether the sound is pleasant or unpleasant.

What is the most unpleasant sound you have ever heard? If you ask a group of students, most will answer that it is the sound of chalk or a fingernail being scraped across the blackboard. The screech makes the flesh crawl, yet it is often no louder than a person speaking.

The difference between the screech and the voice is a difference in pitch, and that is determined by the speed at which each sound travels. When a piece of chalk is scraped across the blackboard or a person speaks, the air begins to vibrate and make waves of sound that move up and down through the air. Some sound waves, like big ocean breakers, must go a long way to climb to a peak, fall, and rise again. If you are standing at the seashore watching such big waves, you have to wait a while between crests. Other waves, like the little ones near shore, go only a short dis-

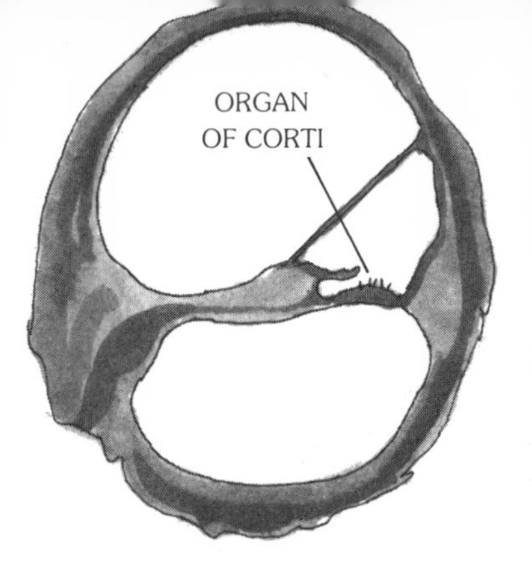

CROSS SECTION OF
THE COCHLEA

THE INNER EAR

EARDRUM

tance to reach each peak. They come quickly, one right after the other.

Although sound waves act like ocean waves, they move much more quickly, reaching one peak and then another hundreds, thousands, tens of thousands, even hundreds of thousands of times in a second. When the wave moves from one peak to the next, it is described as completing one "cycle." Scientists have counted the number of cycles that each type of sound wave can complete in a given amount of time. Those that have a short way to go from peak to peak complete a high number of cycles in a second and are, therefore, the "high-frequency" sounds, more popularly known as high-pitched sounds. Those that have a long way to go from peak to peak get through only a low number of cycles and are the "low-frequency" or low-pitched sounds.

The number of cycles in a second is abbreviated cps or Hz. The Hz is short for hertz, the name of Heinrich Hertz, a nineteenth-century physicist, who was a pioneer in sound research. When one person describes a sound as having a frequency of 1,000 cps and another says it is 1,000 Hz, both are saying the same thing. The term Hz is more often used by scientists.

If you play a musical instrument or sing, and you practice scales, you go from low to high frequencies and back again. The sound wave sent out when you play middle C completes about 262 cycles per second, the A above it 440.

The squeak of chalk is a high-frequency noise, going through many cycles in a second. The screech of a siren and a scream are also high-pitched. You can see, then, that high-pitched noise is often shrill and piercing and is likely to annoy. There are also some pleasant high-frequency sounds, such as the song of the bird or the flute. The low-frequency noise produced by a foghorn or an idling motor may not sound unpleasant, but it makes a strange thing happen. You feel the slow vibrations in your body. Your ears buzz, and after a while there may be sensations in your chest and abdomen.

You hear sounds best when the sound waves complete between 1,000 and 6,000 cycles in a second. That is the range of sounds the human voice produces most often when speaking. A special decibel scale (dBA) has been worked out that gives more weight to these frequencies than it does to higher- or lower-pitched sounds. This scale has become common, and most noise measurements are given in dBA. When you see such a measurement, you know not only the loudness of a sound but also that it is not of either very high or very low frequency.

Some sound waves cannot be heard by the human ear. They may be too slow, fewer than 20 cycles in a second, or too fast, greater than 20,000 Hz. We know they exist, because they have been recorded with special sound-measuring equipment. The very fast sound is called ultrasound, and the very slow, infrasound.

But although humans cannot hear ultrasound, animals can. In the last century the British scientist Sir Francis Galton made a special whistle and surprised his friends with what he could do with it. One day he put the whistle to his lips and blew. No one there could hear the whistle, but a dog came running. A cat that was inside the house began scratching and mewing to get out, because it, too, could hear the high-pitched sounds of Sir Francis's whistle.

The killer whale signals its mate, though we cannot hear the mating call. It is a screech at 120,000 Hz. The bottlenose dolphin can send forth an even more high-pitched cry. When it wants to attract another dolphin, its scream goes out at 270,000 Hz.

chapter three
THE STRANGE WAYS OF SOUND WAVES

A ruler who lived twenty-five hundred years ago in Syracuse, Italy, used to imprison anyone who made him angry. He liked to use a grotto for a prison, because he had discovered that when he stood in a certain spot outside, he could hear every whisper of the prisoners and could prevent plans of escape. The grotto is still called "the Ear of Dionysius," after the name of the ruler, and it is still possible to perform his trick. Sound bounces from the rocks in such a way as to reach the ears of anyone standing in just the right place.

Many centuries later, worshipers in a cathedral in Sicily were startled to hear the voices of people telling their secrets in the confessional. The sound waves of the voices bounced off the stone pillars, hit the walls, and bounced back right at ear level for the congregation. The confessional was moved, and there was privacy once more.

These are just two examples of tricks that can be played by sound. You know many others. Just think of the echo. When sound hits an opposite wall, rock, or mountain, some of it is taken in and some is reflected back, as light is by a mirror. Usually it comes back so quickly— in less than one tenth of a second—that you cannot hear the reflected sound. But when the sound waves must travel a long way before reaching a mountain, the reflection also has to travel a long way back to reach your ear. It travels so slowly that you can hear the exact reflection of the original sound as a second sound. We call that reflection the "echo."

Bats use the echo to find their way around black caves. They give a cry and listen for the reflection of it. The direction from which the sound comes tells them where the opposite wall of the cave must be, and so they can keep from flying into it head-on.

Every time sound waves hit a solid object, like a mountain far away or the wall of a room close by, they bounce. With every bounce a little of the sound is taken in. After a while, nothing is left that can be heard.

A great opera singer liked to impress her friends with a trick. She would place a fragile wineglass on the table in front of her. Then she would sing a loud high note and the wineglass would break. The reason this happened was that the sound wave first pushed the thin glass in and then the wave fell back, pulling the glass back out. This trick is not so easy to copy, which is a relief to those who value their glassware. It can only be done by someone who has spent a lot of time practicing to find exactly the right place to stand and the right note to sing.

You can understand words almost as soon as they are spoken because sound waves travel so quickly through the air. Scientists who study sound tell the story of the man who in 1708 decided to find out just how fast they go. He learned that a cannon was to be

fired in an army drill. And so he found a church that was exactly twelve miles away and climbed its tower. Perched at the top of the tower, he counted the seconds that passed between the time he saw the flash of the cannon being fired and the time the sound of the explosion reached him. By dividing the distance by the time, he got a speed of 1,125 feet per second.

In the twentieth century, experts heard this story. They checked the old figure, expecting to prove it wrong. With the best of modern equipment, they got close to the same result—1,130 feet per second—at a temperature of 68° Fahrenheit. Sound travels more slowly in the cold. During the winter when air temperature is about 32° F., sound can move no faster than 1,087 feet in a second.

Sound not only travels through gases like air, but also through water and solids. It can penetrate walls of stone, brick, or wood. Just go into the living room, close the door and windows, and sit quietly listening. The chances are you will hear the barking dog that belongs to the family next door and the voices of people walking down the street outside. Just how much noise gets in depends a good deal on how solidly built your home is. If the walls are thick, they let in less noise, but any cracks or holes will let sound waves through.

"If an opening is big enough for an ant to get through, it is big enough for noise to get through," says a builder. "After a house has been put up, I walk around it, filling in all the cracks with cement. Both ants and noise can be kept out by careful building."

A local school board was offered some land on which to build an elementary school. The catch was that it was only a mile away from the end of a runway of one of the country's largest airports. The board called in a sound expert to work with the designers and builders. Today you could hear a pin drop in any classroom, provided that the students are quiet. The building is air-conditioned, and the windows are sealed with rubber all around. Some inside walls are four inches thick and others are eight. The concrete roof slab measures five inches. The outside walls are hollow, because air between two layers of brick or board can capture many sound waves.

The walls in many modern buildings, though, are so thin that every sound, from the toilet flushing to a baby's crying, comes through.

"I have the uncomfortable feeling that I know too much about the people next door and they know too much about me," remarks one young homeowner. "My neighbors know just how often the children are fighting or practicing the guitar, and we can hear the television programs that they like best."

Some cities now have rules for builders on just how much noise is allowed inside a house or apartment. These are of little help to people whose houses were built before the limits were set.

Still, no matter how badly a house has been put together, the family living in it can cut down on the noise inside by the way they furnish it. On moving day people are often upset because their conversation and laughter sound so loud and shrill. The problem is that they have not yet put down the rugs or hung curtains. The bare hardwood floors and walls and ceilings do not take in any of the sound waves.

In the Middle Ages, people who lived in great stone castles hung tapestries on the walls to keep out the cold. The tapestries also absorbed the sound of armored feet clanking on the floor and the shouts of the knights. If a modern home is to be a family's castle, thick draperies, sofas and armchairs with fat cushions, and carpeting can do the same thing as the tapestries of old.

Another way of quieting a room is to drown out unpleasant noise with another noise. That may sound strange, but it works when the other noise has just the right mixture of high and low frequencies. This mix is called "white noise," and it can be made with electronic equipment. You may have heard it without knowing that you did. Many doctors and dentists and business offices have white noise made by machinery hidden in the ceiling. If you listen carefully, you can hear a faint hum. You barely notice it, but somehow it covers the noise of typewriters and other people's voices.

You may also hear white noise when you are outdoors. Nature can make white noise every bit as well as an electronic device. Just listen to a waterfall—a low-pitched rumble, a roar, a rushing sound like wind, a high-pitched rustle like that of leaves. It drowns out your voice and the sound of cars from the nearby highway.

Nature has many ways of muffling noise. Just as drapes and rugs absorb sound inside a room, trees, shrubs, and grass take in sound out of doors. The thick shrubbery of a small garden keeps out some of the noise made by people and cars going by. The thicker the shrubbery, the less sound you hear.

What if you were to go to the jungles of Panama where the heavy vines and bushes hide a trail a few feet away? If you were to shout to a friend standing just 100 feet away, you might not be understood. So much of the sound made by your voice would be lost, taken in by the jungle greenery.

But you do not need to travel to the jungle for this to happen. Just go to a place where there are trees and shrubs. When sound is sent across a grove of 150 or more tall leafy trees standing close together, levels are lowered by 5 to 8 decibels. If a forest is really thick, sound may fall by 10 decibels.

A housing project or highway can be made quieter by surrounding it with plants. This is the exact opposite of what most builders have been doing for years. If you have ever watched a highway being built, you have surely seen the bulldozers come in and uproot the trees and plants and cover the ground with pavement. In some places you may also have seen work crews come to plant them again.

Scientists lately have been looking for plants that are particularly good at taking in highway noise. The discovery has been made that corn can absorb a great deal of high-frequency noise of the type made by cars traveling at high speed. If corn is able to do that, so is any other plant with thick leaves, agricultural experts point out. Leafy trees and shrubs planted alongside the highway make life much pleasanter for people with houses nearby.

In some parts of the country the climate and soil are too harsh for leafy plants to grow well. Highway planners can then consider pine trees and hemlock with thin leaves. Even bare brush takes in some of the sound of traffic.

Botanists have also tried to find ways of cutting down the low-frequency roar made by trucks and cars as they start up after stopping at a tollbooth. Plants do not reduce this kind of noise much, but it was found that soft earth does absorb it. A band of soft ground around tollbooths would offer a good solution.

The earth, the trees, and other plant life not only take in noise, but they also make noise less annoying simply by being there. We do not hear with our eyes, but sometimes it almost seems as if we do. An environmental-agency official tells a surprising incident about some people who live in an apartment building close to a busy state highway. The families were made miserable by the noise, and they complained to the city government. City officials went to the state capital again and again to ask that something be done about quieting the highway noise. They were put off repeatedly.

At last the city officials had an idea. They ordered that a single row of trees be planted in front of the apartment house. The few trees made hardly any difference in the amount of noise, but they did block the view of the highway. After that, there were very few complaints from the people in the building.

All five senses work together to form your reaction to the outside world. The beauty of leafy trees can make up in part for the ugliness of noise.

chapter four
ATTACK ON THE EAR

Why is rock music played so loud?

"It surrounds people completely in the rock-and-roll atmosphere. It turns them on."

"The louder it is played, the more control it has over the minds of the listeners."

That is the way a group of students answered a questionnaire about rock. They liked to feel the rhythm and vibrations take them over. Electrically amplified instruments made them feel as if the music were being played right inside their heads. The sound made by a band with electrified guitars is at least as great as that of a jet plane taking off, and it lasts longer. It is exciting, but it is dangerous, too.

A saxophonist, aged nineteen, had played in rock bands since he was fourteen. His hearing was no better than that of his sixty-year-old uncle.

A group of fifteen teen-age rock musicians took hearing tests at the noise-study laboratory of the University of Tennessee. Ten had hearing losses.

Listening to rock very often is almost as bad for the ears as playing it. In Australia a doctor asked young people who heard live rock bands once or twice every week to come in for tests. A large number of them had poor hearing.

"Listening to a rock-and-roll band is one of the greatest risks to hearing that can possibly be found," reported the U.S. Environmental Protection Agency after a careful study. Firing a gun or riding a motorcycle is even noisier. But the careful person will wear earplugs or a helmet for these activities. You never see anyone pull out a pair of ear protectors at a rock concert or at home when records are played as loud as the stereo will go. One rock performer brought his family to a disco; he was very embarrassed when they put their hands over their ears during the best numbers.

After paying for a ticket, many people are annoyed when a band takes long breaks. As noise is most likely to damage hearing when it continues without a stop, those rock musicians who take the most time off between numbers are, in fact, doing themselves and their listeners a great favor.

Listening to classical music is much safer for the ears. Sound experts from San Francisco's Institute of Medical Sciences made a test of a symphony-orchestra performance. They chose "Pictures from an Exhibition," by the Russian composer Moussorgsky, as the test piece. The music was not as loud and the sounds were less piercing than rock. In addition, the frequency range was much wider, with swings from high to low, and that is less tiring to the ear than sounds that are similar in pitch. But more youngsters listen to rock than to Moussorgsky or Bach.

Many people happily bring into their homes noise that could

harm them. Some cap pistols make a louder bang than the rifles used by soldiers in battle or by hunters. A factory owner would have to pay a fine if noise in the shop were allowed to reach the levels heard at home with one television blaring in the living room and another in the bedroom. Snowmobiles are as noisy as a screaming chainsaw. Inspectors from the U.S. Forestry Service visited a number of snowmobile owners and asked how often they drove. Most said that they took their families out for four hours or so on Saturdays and Sundays and "played" with the mobile for an hour or two each day after dinner during the week. Motorcycle riders like to go everywhere on their machines.

Motorcycles, snowmobiles, rock, and cap pistols are often used for fun. Yet the sound that each makes must be added to all the other sounds that we are forced to hear in the course of a day. The Environmental Protection Agency has figured out that a person can stand 70 decibels of noise, as a 24-hour average, without risking hearing. That is at least the noise level on a busy street corner downtown. But many people have much more noise around them. They cannot choose to turn off the sound of the radio playing next door, the neighbor's power mower, the drill making repairs on the street, the trucks carrying supplies down the highway, the subway train, or bus.

Nowadays about one of every ten Americans does not hear perfectly. Noise is not the only reason. Old age, illness, and accidents are other causes. But noise is often to blame. The tiny muscles in the middle ear—the tensor tympani and the stapedius—are strong, and can keep sound out of the inner ear for quite a long time. But when noise of greater than 80 decibels goes on and on, the muscles give way, and the blast of sound forces its way into the cochlea, into the organ of Corti. The tiny hairs, which pass on the sound signals to the brain, bend to the power of the sound waves and snap back.

But sometimes it can take them a while to snap back into place. If you have ever flown in an airplane, you know that you do not hear perfectly for several minutes after landing. The same thing happens when you leave a rock concert or a party where the stereo was blaring. Just notice how your friends raise their voices, because they cannot hear one another at normal speech levels. After your ears have endured the wail of a fire siren, your hearing is poorer than usual, but for a few moments only. On the other hand, just a twenty-minute ride on the New York City subway cuts down hearing for about forty minutes. A student who has gone to a music lesson by subway may not be able to play in tune for a while.

Those hearing losses are not permanent. The hair cells in the organ of Corti snap back into position, and hearing returns to normal. As the years pass and noise is endured more often, it takes longer for normal hearing to be restored. And after a while a hearing loss may come and not go away.

The reason for this is that some of the hairs get worn out by being constantly forced to bend and snap back. In time they bend and stay bent. Some get thin or out of shape, others are completely worn out.

As you might expect, sound reaches the hair cells on the outside rows of the organ of Corti first, and so they are the first to wear out. These are the hairs that carry high-frequency sound to the hearing nerve. That is why people are more likely to become deaf to high- than low-frequency noise. When you speak to them, they cannot hear the "f," "s," "ch," "th," or "sh" sounds. In music they miss the lilt of the flute's melody. It is particularly unlucky that

30
40
VOLUME

fewer hair cells carry high- than low-frequency sound. If one of every five of these is worn out, there is a hearing loss of 40 decibels at high frequencies. What does this mean? If a person with normal hearing could understand you when you talked in an ordinary voice at a sound level of 45 decibels, the person with a 40-decibel loss could hear you only if you raised your voice to an 85-decibel shout.

A hearing loss of 10, 20, 30, or 40 decibels does not happen in a month or a year. And a person does not become completely unable to hear, so long as some hair cells are left. As they are destroyed, one after the other, deafness slowly creeps up. At first it is not very noticeable. Then comes a moment when it is hard for people to follow what is being said to them. They catch only parts of words. The effort to keep up with what friends, relatives, teachers, are saying becomes just too great. Why does everyone mumble so? It is natural to slip comfortably into a daydream.

The more times a person has been subjected to too much noise, the more likely he or she is to become deaf. In addition, hearing, along with all the other body functions, almost always gets weaker in old age. And so many old people are deaf. Patients in a nursing home were tested, and it was learned that of the 181 old people living there, 159 had hearing losses averaging 40 decibels.

Those people were old, but what of the young? Today perfect hearing lasts for a brief period of life. By ninth grade, 9 percent of boys and 3 percent of girls have at least a small hearing loss. The University of Tennessee tested college freshmen for several years in a row. In some classes hearing loss was found in nearly one third of the students, and one year in nearly two thirds.

What is the cause? "I suspect it is because many youngsters go shooting, ride motorcycles, listen to rock and roll," says Dr. David Lipscomb of the university.

While ears look pretty much alike, some can stand noise better than others. Even among rock musicians, you can find a good number with perfect hearing. Hearing tests were given to six thousand men who had spent their entire working lives in a steel mill. Some had good hearing; some were very deaf; others only slightly so. During World War II a number of soldiers became deaf as a result of combat training; others were able to take weeks and months of listening to gunfire. What was the difference between the men? Just as some people have stronger arm muscles than others, so some have stronger ears. It has nothing to do with body size or athletic ability. The strongest person can have weak ears.

Girls and women are more apt to have strong ears than are boys and men. The difference begins to be noticeable at the age of eleven and continues throughout life. At home the mother is more likely to turn down the stereo or television, the father to turn it up.

Some of the sex difference in hearing must still be blamed on the way people live. Women's liberation has not yet caused very many girls to become hunters or pilots. Boys and men are still more likely to go hunting, to serve in the army, fly planes, use power tools, work in construction gangs, serve as subway engineers and bus drivers. But that is not the whole story. Women have stronger muscles in the middle ear. These are the muscles that tighten and keep some of the most dangerous sounds away from the fragile hair cells.

When deafness is caused by noise, it usually comes as a result of this slow wearing down of the hair cells, but once in a while the sound from a violent explosion will kill huge numbers of hairs on the spot, or even break the eardrum or push a bone out of place. Throwing a firecracker near someone's ear is not a joke. It can deafen a person for good.

How can you tell when the noise you are hearing is loud

enough to be dangerous? The explosion of a firecracker or gun is clearly too loud. It makes your ears and head hurt. But it is harder to tell about less startling noises. The most important clue is a ringing in the ears. Even though this usually goes away quickly, it tells you that the noise is too much for you, and you should move away from it. If it takes your hearing a long time to return to normal after you have heard a loud sound, you have sensitive ears and should protect them. If you cannot get away from the sound wear earplugs.

"I wear earplugs when cutting the grass with a power mower," one ear doctor tells patients, giving them a helpful hint.

Putting your fingers in your ears, or even stuffing them with cotton, is not good enough. Protectors must really fit, if they are to keep out the noise of motorcycle, cap pistol, or snowmobile.

But what if people did not have to hear machinery humming, banging, and clattering day after day? What if all we listened to were the splashing of the frogs in a pond and the pattering of raindrops on leaves?

There are still people who live in regions that are quiet, places where the sounds of car, airplane, and rivet are unknown. A few years ago Dr. Samuel Rosen, a New York ear doctor, visited the Mabaan tribe of the southeastern Sudan in Africa. These tribesmen live in a world so silent, he said, that only "the bleat of a goat and other sounds of nature" break the stillness.

Dr. Rosen asked some of the tribesmen to take hearing tests. And what were the results? The hearing of the Mabaan from the age of ten to the age of seventy proved to be far better than that of Americans.

chapter five
WHAT NOISE IS DOING TO YOU

You can close your eyes and shut out light. You can pull your hand away from a slimy worm, spit out food that tastes bad, hold your nose at a repulsive smell. But humans do not have earflaps, so there is no way of shutting out noise. If you cannot run away, you must take it in.

The hero of a spy movie was captured by the enemy and tortured by being tied to a chair and forced to listen to loud high-frequency sound coming at him from all sides.

Hospital patients also cannot escape from noise. An intern measured the level in a room where patients go to recover from surgery. It was close to 70 decibels.

"I have to give pain-killing drugs more often to patients lying in a noisy recovery room or hospital ward than to those in quieter rooms," says a head nurse.

Schoolchildren cannot get away from noise either. And it can have an effect on the amount they learn and the grades they get. Reading tests were taken by students in an elementary school near an elevated subway train track. Every 4.5 minutes a train went by and the noise level on the side of the building facing the track rose to 89 dB. The teacher had to scream or wait for the train to pass. The reading scores of the children whose classrooms were on the noisy side of the building were lower than those on the quiet side.

It is particularly hard to learn something new in a noisy room. Teachers once asked a group of students to take part in a test. Each was given several thin blocks of wood, metal bolts, screws, and nails and told to find a way of putting them all together. Some of the students were allowed to take the things into a quiet room, while the others stayed where it was noisy. One third of the students in the quiet room were able to get all the pieces together in fifteen minutes, while only one eighth of those working against noise did as well.

Even so, you may have noticed that you sometimes get all your arithmetic problems right when working while a softball game is going on in the playground and drivers of cars in a nearby traffic jam are honking their horns. The reason is that you may then be concentrating more deeply on your work. You do not want to hear the noise, so you try very hard to block out the shouts, the conversation, the horns. You can do this for a while, but eventually you will get tired. Even if you are not thinking about the noise, it is always there in the background, trying to break in.

Every time there is a sudden noise—a bell ringing or a plane going by—you jump and lose your train of thought.

There was once a princess who was so sensitive that she could not sleep on a pea, no matter how many mattresses were piled on top of it. Noise is like that pea. And like the princess, we are bothered by it.

You cannot turn off the way your body reacts to sudden sound. Your heart begins to pound and your mouth gets dry. Your muscles tense and your stomach churns. Every time you hear a new noise, you must figure out what it means. Is the door banging in the wind or is someone downstairs? There is a bell ringing. Is it an alarm clock or a fire alarm? And most often, the noise is not a sign of danger at all, but is being made by an angry driver or a helicopter taking off. Regardless, your body acts automatically in just the same way.

This reaction is repeated time after time, day after day. And added to that is the difficulty we all face in trying to shut out background noise in order to work. It is a great strain on the mind and the body. The U.S. Environmental Protection Agency decided to find out whether a great deal of noise could make someone sick. Many people work in factories where machinery clanks, hums, and roars all day long. And so the agency picked some of the workers in noisy and some in quieter factories, and asked them a number of questions about their health. Heart trouble and high blood pressure were found more often among the men and women who worked in the noisy factories.

Loss of appetite in noisy surroundings is common. The food in the school lunchroom may not be quite as bad as the students think it is. The very same dishes might be accepted as fairly good if they were served in a quieter place. Because a sudden loud noise makes the stomach uneasy and the gastric juices dry up, a person may feel slightly sick while eating or directly afterward. A great many people in this country often have stomach upsets or develop ulcers.

On the other hand, the Mabaan tribesmen of the Sudan who live in quiet surroundings and have good hearing also have good health. There is seldom a case of heart disease. In the United

States blood pressure gets higher as people grow older. But a 70-year-old Mabaan has the same blood pressure as a teen-ager. Stomach ulcers are unknown.

After Dr. Rosen came back from the Sudan, he kept track of his Mabaan friends, and so he learned that some members of the tribe had gone to make their fortunes in the big city of Khartoum. Within a few years, many of them began to suffer from heart disease and high blood pressure. Noise is, of course, not the only reason for this. There is more nervous strain in living in a big city than in the peaceful countryside. But noise may have been the last straw.

Even though at one time or another almost everyone says, "That noise is driving me crazy," noise by itself will not have that effect on a sane person. But if someone is very nervous to begin with, then the noise may be more than can be borne.

Psychiatrists at a British mental hospital looked up the home addresses of every patient who had been admitted in the past two years. They discovered that a large number of them lived near a busy airport. Fewer patients came from houses farther away. The psychiatrists reported to a scientific journal that they thought noise was partly to blame.

Even people who would not be driven to the point of mental illness can become nervous, irritable, and ready to pick a quarrel.

"They yell. I yell back. During recess I see a great many fights . . . a rash of brawls. When I ask a youngster what is wrong, the answer is likely to be, 'I feel nervous and jumpy,' " says a teacher from a Boston junior high school near the airport.

In another school, in a quieter place, a student had been looking forward to fifth grade. Her older sister had told her that the teacher was the nicest one in the whole school. "You were wrong," she said to her a few weeks later. "No one can do anything right. The teacher is angry about everything, and she gives the lowest marks of anyone."

One day the teacher mentioned that a new highway had been built near her home. While she corrected homework papers, she heard the trucks and cars rushing by. After she finished, she wanted to relax and read a book or watch television, but the sound bothered her. The change in her personality had been caused by the change in the amount of noise around her.

An old riddle explains what was wrong with her and with the children in the school near the airport. *Riddle:* What kind of a noise annoys an oyster? *Answer:* A noisy noise annoys an oyster, and a noisier noise annoys an oyster more. You could add a line to that and say that a noisier noise at night annoys an oyster most.

"The most harmful noises are those that invade your dreams, because dreaming is necessary for mental health," states psychologist Howard M. Bogard who served on a committee on noise control set up by the mayor of New York City. "Most people have four to six dreams a night. If the dream is interrupted by a jet plane, fire engine, or noisy neighbor, the person will dream twice as much the next night. But if interrupted again and again, this individual will become emotionally upset. And it does not take long. A few nights are enough."

Sometimes a sound will change a pleasant dream into a nightmare. A siren goes off outside, and the dream scene of the happy picnic in the country becomes the nightmare scene of a volcano erupting. You may not remember the nightmare when you wake up, but you know that you did not have a good night's sleep.

A psychologist hired students to take part in a sleep experiment. They went to bed in the laboratory and recordings of noise were played. The sound was kept at a level just too low to awaken the sleepers. When they got up the next morning, they did not re-

member having heard anything. Yet they complained that they felt tired.

Some people are so sensitive that they wake up at the drip of a faucet, while others can sleep through the sound of an electric guitar being played in the next room. In general, though, when the sound level passes 50 decibels, half of those hearing it will begin to toss and turn. That is about as much noise as gets through the walls when a radio is turned on in another room. Most will wake up at a noise that is louder than 70 decibels.

Old people are more easily awakened by noise than the middle-aged, while small children can sleep through all but the loudest sounds. Even the elderly will learn to sleep through a noise if it is repeated often enough. But this does not mean that they—or the young—will ever learn to like it. People do not get used to an unpleasant sound whether it comes at night or during the day. When airplanes keep flying over the house day after day, those who must listen to it get angrier.

Some sounds that are not annoying the first few times they are heard become so when repeated. Many a parent, calm on the first playing of a rock record, is irritated when it goes on again and again.

In the mid-1960's government investigators visited Oklahoma City while air force tests of high-speed planes were going on. They were surprised to discover that as time went by, more citizens complained than at the beginning. Most Oklahoma City citizens never became used to the noise of the planes.

"Noise is most annoying when it keeps you from doing the things you want to do," says an interviewer for a public-opinion poll who went from door to door with a questionnaire. "I learned that people are the angriest when outside noise drowns out a favorite television or radio program."

chapter six
ANIMALS ARE VICTIMS, TOO

If you have ever seen a starling or a blackbird building its nest near the runways of a busy airport, you might think that animals like noise. But the starling and the blackbird put up with noise for a good reason.

"Starlings and blackbirds and certain other kinds of birds do go to airports, because they easily find food there," states Dr. John L. Fletcher, Memphis State University professor who gathered articles on the effects of noise on wildlife for the U.S. Environmental Protection Agency. "But you do not see a great variety of birds. I do not think you will find robins, wrens, or bluebirds around airports."

The robins and the bluebirds are so sensitive that no food can tempt them into a noisy place. You might not expect a skunk to be as fussy as a bluebird, but it is. In the years before the Cape Kennedy Regional Airport was built in Florida, skunks, bobcats, and foxes were often seen. Nowadays these animals are never found near the airport. They have moved to places at least five miles away.

Why should you worry about whether bobcats and skunks have left an airport? The reason is that their action is a sign of how noise made by humans can throw off the balance of nature. Living things depend upon each other. Creatures on earth form a food chain, which is a way of saying that larger animals live on smaller, and smaller animals live on still smaller ones, and so on, until you get down to ones as small as microbes. High-pitched noise from a factory drove insects away from nearby gardens and woods. Homeowners in neighboring communities were pleased at first until they discovered that the robins were hungry. After a while, the robins flew away and did not come back.

Creatures of the wild are in the greatest danger. Year by year machines invade more of the wilderness. Snowmobiles now run over the coldest parts of the world, passing across ice-covered land that once was home to polar animals only. Food is scarce in these cold places, and many creatures manage by sleeping through the winter months in a cave or den. While they sleep, or hibernate, as this deep winter sleep is known, they need not eat. They come out of their dens when spring and warmth have brought the plants to life again.

What happens to a hibernating bear or Arctic squirrel when a snowmobile rushes past on a winter's day? If the hibernator woke up and came out of its den too soon, it would starve. We are not sure that this is really happening right now. It is hard for scientists to find the dens of hibernating bears and squirrels and keep watch until a snowmobile goes by. But thoughtful people are worried.

We already know what happens to many wild animals when they hear noise. The Laplanders who live in the far north of Scan-

dinavia let reindeer out of their pens during thunderstorms. "The thunderclap makes them frantic. They stampede and, if crowded together, knock one another over," explains a rancher.

Airplanes and machinery are as terrifying to animals as the thunderclap. Before the building of the Alaska oil pipeline, a wildlife experiment was carried out. Loudspeakers were set up to broadcast the kind and amount of noise the machinery running the pipeline would make, 108 to 111 decibels, at a distance of 15 feet. Rams stayed a full mile away from the noise, and caribou would not come within one eighth of a mile of it. Flocks of snow geese flew in a great detour away from the equipment.

Caribou, moose, even bold wolves and bears, ran or trotted away from low-flying airplanes and helicopters. Airplanes flying high in the sky frightened the more timid animals. Sheep went into a panic.

In trying to get away from noise, animals could be driven out of their nesting or mating grounds. They would have fewer young.

We value the bald eagle as the American national symbol. This bird does not like noise. One year a team of scientists counted the eggs in the nests of bald eagles on a mountaintop. They came back the next year and found only half the number of eggs. Many of the nests were empty. What had happened? During that year, helicopters had flown overhead and disturbed the birds.

For many years 30,000 sooty terns, the seabirds of Florida, were hatched every season. Then one year hardly any babies were born. That year low-flying, high-speed military planes had passed over the nest area. A National Park Service biologist blames the airplanes for frightening the mother birds away from the eggs.

Some kinds of birds are already very scarce. They might disappear from the face of the earth altogether, if too many mother birds were frightened from their nests.

"I am worried that noise is not even mentioned when an animal is known to be so scarce," states Dr. Fletcher.

Noise may not only push animals out of their breeding grounds, but may even make it harder for them to mate at all. Sound has a sexual meaning for animals. A male mosquito tried to mate with a tuning fork that was giving out the tone normally used by a female wishing to attract him. A female then rushed to a loudspeaker playing a recording of the sex signal of the male. At that very moment a male insect was right nearby, but she paid no attention to him because he was quiet.

Many different species or kinds of frog live in the same forest pool. Yet any female can find the right partner by recognizing his call. Each species of frog broadcasts its mating call at a particular frequency.

A turkey can pick her own chicks out of a flock of young birds that look alike by hearing their cheeping. If this is drowned out, she does not know which chicks to take care of.

Birds send complicated messages warning of the presence of an enemy or telling where food is to be found. Bees give out short bursts of sound when there is danger. The gorilla beats its breast, the wood pigeon flaps its wings, and the stork snaps and clatters its bill. Even the earthworm signals by hitting its body against the ground.

These creatures can be confused when a machine, car, airplane, bell, or siren just happens to send out a sound like the mating cry or warning signal of the bird or the frog. Some machinery gives out sounds at frequencies too high for the human ear, but insects can easily hear them and become bewildered.

During World War II, submarine captains suddenly became interested in the snapping shrimp. This little shellfish makes a loud crack every time it closes a large claw. The sound, the submariners learned, was picked up by sonar. This is a system which sends out electronic signals that bounce off rocks or other submarines, telling where they are. The shrimp noises confused the humans into making wrong moves.

Four blue whales were swimming one day underwater near the coast of Chile. The sounds they made were recorded by scientists in a submerged research chamber. "The underwater moanings were the most powerful sound recorded for animals—188 decibels at a distance of three feet," reported the naval oceanographers.

Then they added the fact that the sound is at "the same overall noise level as that of a United States Navy cruiser traveling at normal speed."

Noise warns animals, as it does humans, that they are in danger. They, too, are alerted to fight or to run away each time they hear the sound from a cruiser, airplane, or automobile. This reaction can be bad for animals that are nervous by nature. The animals themselves cannot complain, but the farmers who own them often do. Many lawsuits have been brought by farmers against building or airplane companies or the air force for frightening animals so that they could not give milk, or lay eggs, or care for their young.

As one farmer pointed out, "Animals are not given to lying or refusing to work or give milk. So it should be easy to prove the case against noise."

But many courts disagree with this, seeming to believe that somehow the animals are to blame for not telling how they feel about noise in words. And so some cases are won by the farmers; others are lost. Norwegian fur farmers came to court to complain that a mother fox had eaten her young when there was loud blasting nearby. The court blamed the company that had done the blasting. One mink breeder in England was given several thousand dollars when he showed that the flights of high-speed planes had turned his mink into "nervous wrecks." But when an American study was made of other mink in Alaska, they seemed to be less sensitive than the British animals. The mink behavior was watched while military planes from the Elmendorf Air Force Base in Anchorage flew overhead. The noise made by the planes was so great that it shook the earth three times in an hour and a half. "By the third [time], the mink were not even looking up," said a witness.

There is the same kind of confusion about how plants react to noise. Some plant lovers are certain they are damaged, while others believe they are left unharmed. One woman played recordings of music to her plants. When East Indian temple music came over the loudspeaker, petunias, carnations, geraniums, beans, and squash burst into bloom. But hard rock music made them droop, and in three weeks they were dead. A violinist went out into his garden and played sonatas every day. It seemed to him that the flowers were more beautiful than before. Farmers in the rice paddies of Madras, India, played the flute for half an hour a day and then insisted that the rice plants grew to be taller than normal. A professor blasted a leafy plant with sound as loud as that made by a train pulling out of a station. The plant wilted. The explanation given was that plants respond to loud noise by losing water.

We cannot say that any of these people really proved that plants react to sound. There have been too few recorded experiments so far. Many more tests must be performed before we will know if music really does make plants grow better and noise makes them wither. Do flowers in the garden of a music school bloom more fully than those in the yard of a steel mill? We do not know.

We do know, however, that plants cannot stand air pollution and tend to be small and spindly in places where there is a lot of smoke in the air. Could noise finish off those plants that are already half dead from bad air? Perhaps someday we will have the answer.

Plants have an important place in the food chain, and if many of them were to grow poorly, some animals, including some humans, could go hungry.

chapter seven

THE LAWS OF SILENCE

In the old days when cities and suburbs were not so large and crowded, most people enjoyed fairly quiet lives. They were very annoyed if any loud noise drowned out their conversation or woke them up at night. So many complaints were made to the mayors and governors whenever this happened that laws against noise were passed.

The things that annoyed people most at that time were loud shouting or fights in the street or a dog barking for hours late at night. And so these were the noise-making activities that were banned by law. Most cities and small towns had such laws, and people began to call them "the barking-dog laws."

The years passed, and cars and trucks in great numbers raced down the highways, airplanes flew overhead, and building crews drilled holes in the streets to lay down power lines or pipes for sewage. Barking dogs came to be the least of the noise problems. But for a long time no better laws were passed.

Then forty or more years ago a newspaper editor in Memphis, Tennessee, became ill and had to stay home from work. Every evening a girl in the house across the street from him had a date. Her boyfriend would get there early, park his car at the curb, and keep honking his horn until she came out. The first few nights this seemed funny to the editor; after that, he began to be angry.

When he recovered from his illness, he walked around Memphis listening and watching, and he discovered that many drivers, like the young man, blew their horns when slightly cross or impatient. Yet the only good reason for blowing a horn is to warn other drivers and people on foot that a car is coming.

The editor wrote newspaper articles asking city officials to do something to stop the noise. People living in Memphis read the articles and began to talk about them. Schoolchildren asked their parents. After a while, drivers began to feel embarrassed about blowing their horns. In 1938 the government decided to make it against the law for anyone to honk an automobile horn, except as a warning. By 1940 Memphis had become known throughout the United States as the "quietest city."

One city and state after another decided to follow Memphis's example. Then each one added to the old "barking dog" laws the things that bothered its citizens most. It became illegal to keep a rooster within the city limits of Chicago, to be drunk, noisy, and disorderly in Pennsylvania, to have a merry-go-round sending out cheerful tunes near private homes in Texas, to play a radio without an earphone while riding on a bus, subway, or ferry in New York City.

In some states and cities the laws declared that cars, drills, and rivets should not make as much noise as they do. But somehow the cities and suburbs kept getting noisier. More laws were passed, forbidding any person or machine from making a noise that is "too loud, unreasonable, or unnecessary." These did not help much. Nobody was quite sure how loud "too loud" really was and so could not be sure whether they were really breaking the law.

And so the lawmakers decided to try something new. They would figure out just how many decibels of sound are "too many" for a person, or a car, or a truck to make. As a start, a group of men and women volunteers took tape recorders to the side of a busy highway and recorded the sound made by the trucks going by. Many of the sound levels turned out to be higher than 87 decibels and sometimes went over 90 decibels. Noise that loud drowns out conversation and television, and if heard for too long, can cause some loss of hearing.

Upon learning this, officials in the state governments of New York and California set limits on the number of decibels of sound that trucks and cars would be allowed to make. If they were louder than that, they could not use any of the state highways. This meant that manufacturers would have to make changes in the engines, and truck and car owners would need to get good mufflers and to fix anything that rattled or squeaked.

It did not take long for people to realize that a law passed by just a couple of states could not really work. Trucks made in one state ride down the roads of all the others. The manufacturers would not make special cars and trucks just for New York and California when their old noisy ones were accepted everywhere else.

And so citizens of many parts of the country began to write to their congressmen and senators, saying that the only way to make the highways and streets of cities and suburbs everywhere quieter would be to pass a national law.

If you were to take a poll and ask Americans when the national

Noise Control Act was passed, almost everyone would answer that this must have been done fifty, or a hundred, or even two hundred years ago. After all, Benjamin Franklin complained about his sleep being broken by horses and carts passing under his windows. But no law was passed in Franklin's day, or for many, many years afterward.

The Noise Control Act was passed by Congress in October 1972. And it was bitterly fought to the last second by groups that were certainly in favor of cutting down noise—only not the noise of the particular piece of machinery they were making. Finally they gave up, and the law went through. It ordered that studies of noisy machinery be made and citizens be given a chance to speak out at public hearings. Then manufacturers would be told just how much noise in decibels each piece of equipment would be allowed to make.

"At last we are telling people what 'too loud' really means," said a government lawyer.

You may not yet have noticed very much of a change. In April 1956, decibel readings were taken at six different places in New York City. These measurements were written down in a book which was placed in a bank vault along with other facts about life in the city at that time. Twenty years later the vault was opened, and city environmental officials rushed to the same six places with their sound-measuring equipment. Five of the six were exactly as noisy as they had been twenty years earlier. Yet this was four years after the national noise law was passed. Other cities and suburbs are still noisy, too.

It will take time for the effects of the law to be felt. The U.S. Environmental Protection Agency had to pick out the noisiest machines and ask engineers for advice on how much each could be quieted and for how much money. The agency then began to set noise standards in decibels for one product after the other, starting with trucks. The rules were announced only in 1974, and truck manufacturers had until 1978 to make the first changes.

When all the noisiest machines are finally quieted, the level of noise in the world around you should not be a danger to hearing and health; it should be safe for both the body and the mind.

Have you ever wondered what happens to the hearing and health of the people who work in the factories where the trucks, cars, or airplanes are made, in the steel and rubber mills, in the chemical plants? If you have ever visited a factory, you know how noisy it is. The machinery that is used to make the car or automobile engine will clank, jangle, hum, roar, creak, and screech. Many different machines, all of them noisy, are going at the same time. Day by day, year by year, the sound batters the delicate hearing organs of those nearby.

"I always had to shout to make my father hear; he worked in a steel mill," said an old man, remembering his boyhood in the early 1900s.

More recently, in the 1960s, hearing tests were performed on men in an engine room. About 15 percent of them were deaf.

Yet owners of factories have been slow to admit that noise heard during the working day was the cause. One of the speakers at a British meeting on noise said sadly that "child labor in its heyday was not considered an evil practice by parents or employer. In a new form this attitude continues to be present, this time about noise."

"A man who takes pride in his work could work in a lumber mill indefinitely without harm to his ears," a committee of lumbermen told a university professor who had come to discuss conditions with them.

Some workers have not been willing to accept their loss of hearing without fighting back. In 1928 a young steelworker named Matthew Slawinski got a job with the J. H. Williams Company. Over the next twenty years, deafness crept up on him. Slawinski was still able to hear well enough to do his work, but he was unhappy. He could not easily hear what his friends said or listen to the radio. By 1948 he was angry enough to go to court. The company was found guilty. This was the first time that such a thing had ever happened.

"Even if a man can go on working, but cannot enjoy listening to music, he has been wronged," explained one judge.

Slawinski was awarded only $1,661.25, which, you can see, is very little money for a big company to pay. Even so, many factory owners stopped to think what the case might mean to them. Surely many workers in their plants were also losing their hearing and might decide to follow Slawinski's example. Most employers decided that they would rather pay a thousand or so dollars every now and again than to spend a lot more money and take the trouble to buy quieter machinery and to soundproof the ceilings and walls.

But the employees kept on insisting that something must be done for them, and in 1969 the government decided that any company with government contracts had to limit factory noise. In 1971, the Occupational Safety and Health Act, covering all other businesses, was passed. This act stated that no person was to spend a full eight-hour working day surrounded by noise louder than 90 decibels. No more than 15 minutes in any day may be spent at a machine putting out 115 decibels. If workers in a plant must hear louder noise for longer periods of time than is allowed by the act, the employer must try to find ways of improving the machinery. And in the meantime, ear protectors must be given out.

Ear doctors and environmental officials immediately objected to

90 decibels for an eight-hour day, and began to work vigorously to lower the limit. Their point was that even 85 decibels is as loud as the sound you hear when standing on a downtown street corner at an hour when schools and offices are letting out.

Nonetheless, the law from the start helped workers from many noisy industries. Last year the foreman of a textile mill complained to the local environmental agency, and an investigator was sent to check on the noise. He walked into the weaving room when the looms were going at full speed and measured sound of 105 decibels. The manager was given a warning, and quickly added foam rubber and other soundproofing materials to the old machines and bought some new quieter ones. In just a few weeks the noise in that room was down to the legal level.

"Now we can talk to each other again," said one of the workers.

chapter eight
THE NOISIEST WAY TO TRAVEL

On a December day in 1903, Orville Wright and his brother Wilbur made their first flights near Kitty Hawk, North Carolina. The longest flight took slightly less than a minute, but the air age had begun.

In those exciting days noise must have seemed a small price to pay for the miracle of flight. Humans had long dreamed of being able to fly like a bird and skim high above the earth. There is an old Greek myth that tells how a young man, Icarus, fastened to his back enormous wings of feathers held together with wax. They carried him high into the air, but the wax of his wings melted in the sun's rays, and he fell.

The winged machines made in the early years of this century did keep humans in the air, and the early flights must have seemed as wonderful to the people living then as spaceships traveling to other planets do to us today. Now, as then, few people mention the noise produced by the scientific miracle. Yet a rocket blast-off makes 180 decibels of sound.

By now people take flying for granted and are tired of airplane noise. You may hardly ever take a trip by plane, but you must listen to planes flying overhead anyway. Some high school students in a town near an airport decided to count the number of flights going by. They used the spire of the church as a landmark and took turns watching the planes fly over. This happened once a minute during the daytime hours.

"I like it here when the airport is fogged in and planes cannot land," said one of the students. "It's such a relief."

About twenty million people in the United States live near enough to airports to want that relief. In 1969, sixty-six years after the first minute of flight at Kitty Hawk, the first federal law ordering that airplanes be made quieter was passed. Manufacturers were told that new planes had to make less noise. They could produce no more than 108 decibels of sound at takeoff or landing.

"The soundproofing material we put on the 747 weighs as much as twenty-six passengers," says one airline official.

Changes in older planes, built before 1969, have also been ordered. They must fly more quietly or be grounded.

You probably hear about "whisper planes" on television commercials, but if you think planes are still noisy, you are right.

"We would like airplane makers to cut airplane noise by 10 decibels every ten years until engines really do 'whisper,' " say government engineers.

With this in mind, a new engine has already been designed. It is called the "quiet engine," because it makes only 90 decibels at takeoff. This is quiet for an airplane, noisy for almost anything else.

"If you are wondering why no one is using the quiet engine," remarks one of the engineers, "it is just too heavy. The airplane can't get off the ground. We need a few more years to solve the problem."

One way to let the millions of people who live near airports get their sleep in the meantime is to forbid planes to land or take off late at night. Such a ban is more common in Europe and Asia than here. If a plane reaches Japan's Osaka Airport after a certain hour, the pilot is told to fly on and find another airport where the rules are less strict. Sometimes this means flying for hours. A pilot will not make that same mistake twice.

There are a few flying tricks that can help in the daytime, too. The pilot keeps the plane high in the air as long as possible, and then comes in for landing at a sharp angle. It is even harder to be quiet at takeoff than at landing, because the plane needs all its engine power to lift it up. But as soon as the plane is off the ground, the pilot can cut back the engines, and so cut down the noise. The power is kept low while the plane flies over houses, apartment buildings, and schools. When it gets to more deserted land or the ocean, the pilot turns on full power and climbs higher still.

The very small aircraft, the helicopters, take off in a blast of noise. The propellers have to turn violently and noisily to get the helicopter up without a long warm-up run along the ground. They must keep spinning with great force to keep the small plane aloft.

Their noise annoys a great many people, because helicopter airports are often placed right in the middle of a busy city or suburb. Some helicopters must be kept handy for the police to use on rescue missions—to find a child lost in the woods, to rush someone suddenly taken ill to a hospital, or to locate a traffic accident.

But sometimes helicopters are used just to take a few people to their jobs in a nearby city.

A big corporation had a helicopter airport built in a quiet suburb for this very reason. "Take away the heliport," the families living nearby told the management angrily. "Why should everyone suffer so that a few people can get to work faster?"

The same question is being asked a great deal nowadays, because a new plane, which is particularly noisy, has come into use. You have probably heard people arguing that the British-French airplane, the Concorde (pronounced Kon'-kord), should not have been allowed to use American airports. This plane can fly through the air faster than sound waves can travel. It is called a supersonic transport plane, or SST (super means "above"; sonic, "the speed of sound"). The Concorde has been getting the most publicity here, but the Soviet Union has a supersonic plane, the Tupolev-144, and there are also many American and foreign military supersonic planes.

Just how fast must a plane fly to go faster than sound waves? The answer is more than 660 miles an hour when 35,000 feet up in the air. (Sound waves travel at different speeds, depending on how high they are.) You will hear the speed of sound described as Mach 1 (pronounced Mack) after the name of an Austrian scientist, Ernst Mach, who studied sound. At Mach 2 the plane goes twice as fast as sound, or 1,320 miles an hour. An army plane succeeded in flying at Mach 3. Going at three times the speed of sound, it crossed the Atlantic Ocean in just over one hour and fifty-five minutes.

Although Ernst Mach died in 1916, long before there were any supersonic planes, his experiments showed what could happen should an object move through the air that fast. As the plane rushes along at supersonic speed, the air is pushed into waves behind and beneath it. These hit the earth with a bang, called the sonic boom, that sounds and feels like an explosion a block away. People living miles away feel the plane going by, even though they cannot see it. Windows rattle, and dishes shake on cabinet shelves.

People used to think the sonic boom was interesting. In the late 1940s and the 1950s, pilots made such booms to order as a featured act in air shows. No one then dreamed that the sonic boom could become a part of daily life.

One morning in late summer a Navajo Indian was walking near the Canyon de Chelly in Arizona. All at once he heard a bang and saw a big piece of overhanging cliff fall off. While he watched in horror, one boom after another followed, and each time some more of the rock face broke. Ancient people had carved homes out of the cliff thousands and thousands of years ago. As the supersonic planes went by, these prehistoric shelters fell to the bottom and were shattered.

Across the ocean, in France, a supersonic plane flew over a wooden farmhouse just as a farmer, his wife, and eight workers were sitting down to lunch. The sonic boom shook the old building, and the roof beams collapsed. Eight tons of barley, which had been stored in the attic, fell down and buried the ten people below. Three of them were killed, and a fourth was badly hurt.

The U.S. Federal Aviation Administration has passed a rule ordering supersonic planes to slow down to normal airplane speed when flying over land. They can travel at supersonic speed only when going across the ocean. While this rule keeps the sonic boom away from places where people live, it does not protect the living creatures in the ocean. When thirty-nine biologists and ocean experts at the Woods Hole Oceanographic Institute heard about this, they were upset. No one has studied whether sonic booms may harm sea life, they warned.

Engineers talk of finding ways to prevent the boom. Some say that if the plane could be improved so that it could fly at a height of more than 100,000 feet, the sound would not reach the

ground. Others say that sonic boom would not be felt if the plane flew at a height of 60,000 feet, but went just a little faster than the speed of sound—up to Mach 1.15. And a third group of scientists is claiming that if the plane could fly at six or more times the speed of sound, the sonic boom would be less strong.

But people living near airports where supersonic planes are landing and taking off point out that the SST does not need to fly faster than sound to be very unpleasant. At any speed, the engines are extremely noisy and annoying. A *New York Times* reporter, watching a supersonic plane take off, declared that "with engines keening like a banshee," it taxied down the runway.

On a summer's day a few years ago, the Concorde set off from Paris on its first flight to the United States. As it flew over the ocean, it went faster and faster, reaching Mach 2. It arrived in Boston, having flown 3,060 miles in three hours and ten minutes. But a young woman who was meeting another plane that arrived a few minutes later, shook her head. "The Constitution of the United States includes the words 'to ensure domestic tranquility,' " she said. "To my mind, that includes quiet."

Since that day Concorde flights have become almost commonplace, yet the problem of the noise they bring is no closer to being solved. Jet airplanes, both old and new, are still much too noisy, but at least they can meet the noise standards being set by the government. The Concorde is an exception. As the U.S. Federal Aviation Administration points out, there is today no known way of making the supersonic planes as quiet as the jets.

The supersonic plane makes it possible to travel faster than ever before, faster than your grandparents believed humans could ever go. But against the advantage of great speed is the disadvantage of great noise.

chapter nine
SEARCH FOR A QUIET MACHINE

"We have the engineering ability right now to quiet the truck, motorcycle, motor boat, air conditioner, vacuum cleaner, toilet, washing machine, and any other appliance or machine you can think of," says a noise expert, Lewis S. Goodfriend.

Instead, how much noise do you hear at home when the appliances are running? If you have ever tried to make a telephone call while the vacuum cleaner is on, you know how hard it is to hear the voice on the other end. Dr. Lee E. Farr of the University of Texas Graduate School of the Biomedical Sciences decided to check out the noise in an apartment in Houston. It was on a quiet street, but even so the sound coming in through the windows was 50 decibels. Then the vacuum cleaner was plugged in and run over the rugs and furniture. The hum brought the noise level in the living room up to 81 decibels.

Dr. Farr checked the kitchen before and after dinner. The family, he learned, had as much, sometimes more noise around them than a worker on the assembly line in a factory. When the vent fan over the stove was turned on to get rid of the cooking smell, the sound level reached 84 decibels. After dinner when the dishwasher went to work, 88 decibels of sound were heard in the kitchen. The leftovers were tossed into the sink for the garbage-disposal unit to grind. The disposal began at 91 decibels and hit peaks of more than 100 when grinding some particularly tough cobs of corn. That is the quiet life at home.

A manufacturer was asked why he did not make quieter dishwashers and washing machines when this was possible. He answered: "But no one asks for them."

When vacuum cleaners, washing machines, cars and trucks, were first invented, they hummed and rattled. In those days manufacturers did not know how to make quiet machines. Boys and girls grew up to the loud hum of the vacuum cleaner, the clank of the washing machine, and the roar of the automobile engine. If they complained, they were told that a powerful machine has to be noisy. When they grew up and bought cars and appliances of their own, they wanted power, and they looked for noisy machines.

Nowadays it is possible for a quiet machine to be as strong as a noisy one, but many people do not believe it. A manufacturer of a power lawn mower recently decided to redesign the motor of the machine so as to cut greatly down on the sound. Nothing else about the mower was changed. After a short time many of the new lawn mowers were returned to the stores by angry customers. They were sure that the machines were no good. Even though the grass had been cut as well as ever, they insisted that the new machine was less powerful than their old one because it was so quiet.

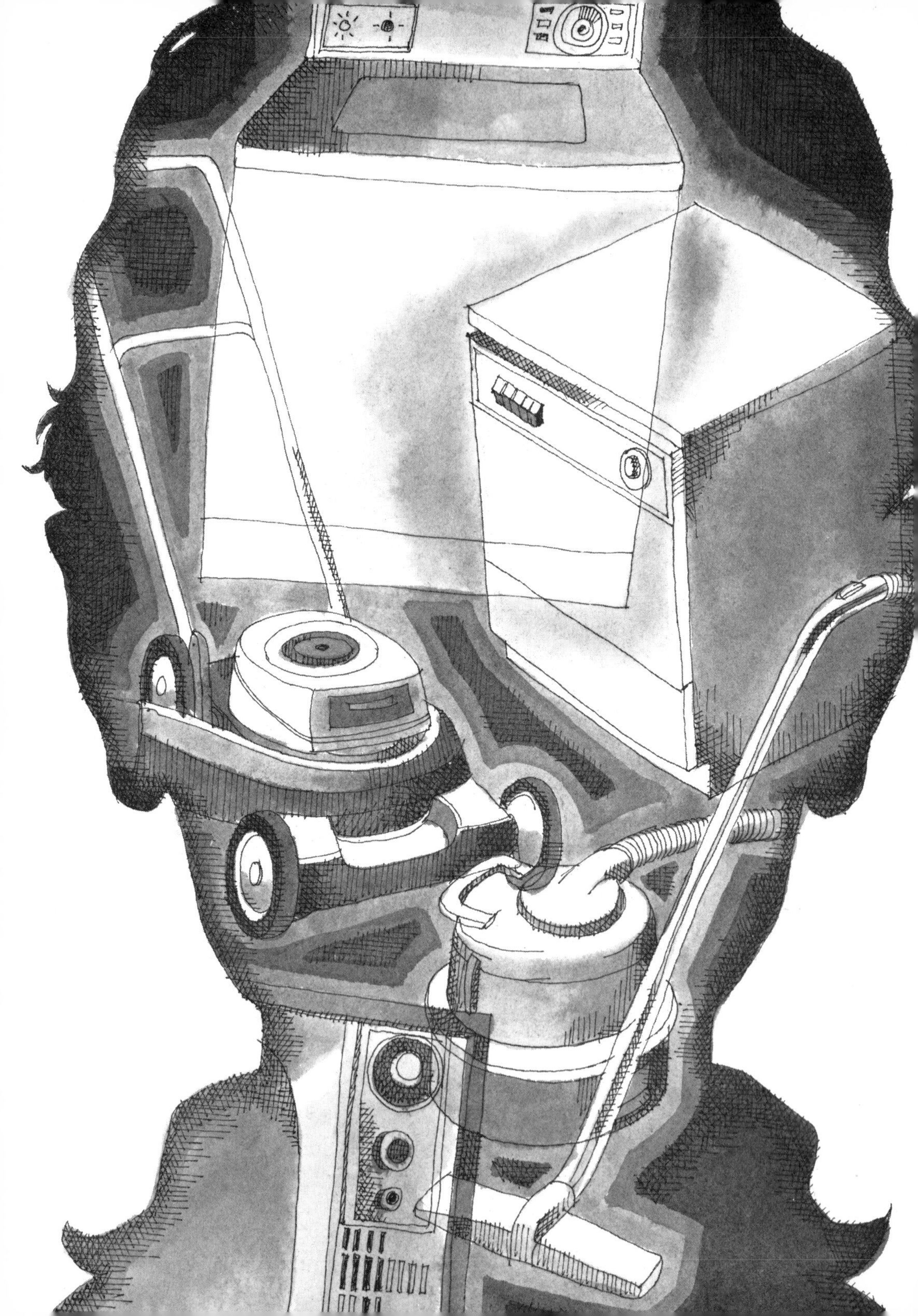

The U.S. Environmental Protection Agency studied the advertisements of appliances in hundreds of mail-order catalogs. Only one of the many advertisements for washing machines used the word "quiet" as a selling point.

Yet when quiet is considered important, machinery is hushed. Because any sound might be heard by an enemy, equipment used by the army runs almost silently. A teletype machine or a generator that makes electricity cannot be heard six feet away.

"Most of the designs used for military goods have been around for years," says Lewis Goodfriend, "but the ideas have not been taken over for civilian goods."

The president of a company making automotive machinery was told that one of his machines was much noisier than those of other companies. He called in engineers and asked them to quiet the machine. The engineers worked for months and finally came up with an engine that gave out 22 decibels less of sound.

"That is too quiet," complained the manufacturer. "We can meet the competition if you bring the sound down by 4 decibels. There is no need to spend the money to make machinery quieter than that."

"When customers insist on quiet machines, companies will make them. It is as simple as that," declares an environmental agency official. "If that should happen, suddenly, as if by magic, engines will have mufflers. Gears and wheels will be surrounded by plastic and foam rubber and other soundproofing material. Soft plastic parts will be used instead of hard metal. When these hit against one another, little sound will be heard. Machines that make noise because they shake as they run will be put on stands made of soft rubber."

Some manufacturers are coming up with imaginative ways of making things quieter. One company, for example, has invented a soundproofing paint. When two bars covered with this paint are struck together, only a soft thud can be heard.

Even without a change in the materials used, most machines could be run more quietly. The easiest way to cut down on car and truck noise, for example, is to lower the speed limit. This has the obvious added advantage of saving lives. The faster a car goes, the more likely it is to have an accident; the faster a car goes, the more noise it makes. If you travel to Canada, you may go past a big thermometer by the side of a road. It takes a car's speed "temperature." If the car is going too fast, the thermometer shows a fever.

Cars are supposed to have mufflers, but some drivers do not like them. Policemen stopped a hot rod on the road one day and brought it to a garage. The driver insisted that the car had a good muffler. Considering how much noise it was making, this did not seem very likely, so the suspicious policeman took out the muffler and cut it in half. The insides had been removed, and the cover put back. The driver's license was taken away for a while.

The U.S. Environmental Protection Agency is going down its list of "noisiest" machinery in use and is ordering big changes made. Trucks, building equipment, motorcycles, buses, airplanes, trains, power mowers, are all on the list. In some cases the changes may be slow in coming. The agency says that it will take about twenty-five years before all the old locomotives are retired to train graveyards and replaced with new quieter ones.

There are a few machines that cannot be quieted no matter how hard the manufacturers try. Even then, the people who must be near them can be protected from the noise. One of the loudest sounds created by humans is that from the blast-off of a Saturn V rocket. This noise is loud enough to break the eardrums of anyone standing nearby with ears uncovered. Yet it is necessary for the people at the Ground Control Center at Cape Kennedy to be close by and see what is happening when a space shot is set off. So that they can do this without being harmed, the launch control center has windows that are made like a five-layer sandwich. A specially developed soundproofing plastic has been placed between each layer.

Not many of us will be that close to a space shot, but more than 27 million Americans must listen to the noise of buildings going up or being torn down. The U.S. Environmental Protection Agency has set noise limits for some of the equipment, but there are other ways of protecting people, too. Several years ago a fifty-two-story building was to be put up in the middle of a busy city. The people living nearby were dreading the blasting. But when the time came, all they heard was the whistle signaling that the blasting was to begin. Steel wire-mesh blankets spread over the ground kept in the sound of the blasting. In addition, the building had been designed so that the metal framework was fitted together without rivets.

This kind of building would become more common if people complained about noise the way they do about bad air or dirty water. You have probably heard a lot about the need to protect the world from pollution. But you may never have heard noise mentioned as a kind of pollution.

Noise is harder to pin down than air or water pollution. Nothing is left after a noise. It was here and is gone. Was it annoying? Did it make people irritable and tense? Even hearing loss comes on so slowly that it is hard to tie to noise.

And so noise pollution is less dramatic than air and water pollution. And yet we must listen to the sound of building crews drilling and riveting, to horns blaring and tires screeching, to airplanes flying overhead. When all of these go on at the same time—and they often do—the moment has come to recognize the dangers of noise pollution. The air around us should be cleared of noise as well as of dust and smoke.

chapter ten

YOU CAN SAY NO TO NOISE

It started with just one person. The traffic noise on the street in Boulder, Colorado, where this man lived got louder year by year. When he called the mayor's office, he was told that Boulder had no noise-control law. He decided to bring together a group of people who would work to get such a law passed. Students from the city's high schools and from Colorado University and businessmen and engineers formed a committee.

They thought their best chance lay in drawing up the kind of law they wanted and then bringing it to the city council. It took them a year and a half of studying law to be able to do this. When the members of the city council saw this fine law, they passed it. That was in January 1970.

The committee wanted to keep students interested in noise control and so they asked experts to talk at the high schools and university.

"We give each student at Colorado University (and there are about twenty thousand of them) an information sheet on the noise-control program when he or she signs up for a parking sticker for the campus," said Boulder's environmental-protection officer. Few students break the law after reading how important it is.

People in Boulder notice noise makers and report them. One student heard an unmuffled car roar by. She managed to jot down the license number, and sent it to the environmental agency. The owner of the car was ordered to pay a fine . . . and to buy a muffler right away.

The example of the citizens of Boulder was followed by people in many other cities and small towns. Some had no laws; others had laws so out-of-date that they did no good. These were the old "barking dog laws" that did not mention traffic. The U.S. Environmental Protection Agency got so many calls and letters asking for advice that local government officials asked a group of lawyers, the National Institute of Municipal Law Officers, to draw up a sample law for communities to use as a guide in drawing up their own law. It is called the "Model Community Noise Control Ordinance," and it is put out by the U.S. Environmental Protection Agency (Washington, D.C. 20460). Nowadays, with the help of this model, a committee need not spend a year and a half in legal studying before presenting a bill to the city council.

Even when there is a good law, citizens must insist that it be kept. Complaints about noise can bring results. A homeowner in a suburb near a large and busy airport was very angry about the airplane noise. He wanted to make his complaint so startling that everyone would have to pay attention. He went to the library and took out a book with pictures and descriptions of weapons used by the knights in the Middle Ages. He found information about a slingshot that had been used to hurl stones over castle walls. On the basis of the old plans, he made a huge slingshot. Then he had to figure out what to shoot. He did not want to hurt anyone, just to get into the newspapers. And at last he had an idea. He baked several batches of muffins. Then when the planes flew over his house, he shot the muffins at them with the slingshot.

Most people object to noise in a more sensible way. They write to their city governments. A few years ago officials in the health departments of a number of different cities looked over all the letters of complaint to find out who had sent them. In this way they learned that rich people complain more often than poor; well-educated people complain more than those who did not go to school for very long; and old people complain more than young ones.

Young people are beginning to join in the noise battle, too. In New York City, an antinoise citizens' group decided to set aside one week in the year as Quiet Week. During that time everyone was to think of ways to make the city more quiet. Second and third graders painted pictures showing first a quiet city and then a noisy one. A friendly store manager put the pictures in the front windows of her store. So many people came to that little store to see the paintings that the director of an art museum asked if he could borrow them for a show.

Children in an elementary school near Japan's Osaka Airport wrote compositions telling how they felt about the planes going by. The articles were so touching that a Japanese antinoise group raised money to publish them. If you ever go to Japan, you might ask for that book. It is called *No tobanaide,* which means in Japanese "don't fly anymore." Many American schoolchildren feel the same way.

Until a new highway was built, it was possible for teachers to teach and students to learn in an elementary school in Elizabeth, New Jersey. The highway traffic raised the noise level in the classrooms by 20 decibels. After getting a great many complaints, the city's Board of Education decided to check. They asked that tape recordings be made inside the classrooms. When these were played in court, the judge said that no student could be expected to learn anything in a room where such noise was going on, and ordered that the school be soundproofed.

A disco opened on a block that had been quiet. A neighbor told the owner, "We'll only be happy when you tone down the music." The families living there joined a block association. They hired a lawyer to present the case against the disco in court. The owner of the disco was surprised that high school and college students living on the block objected, too. But many people who enjoy rock music are beginning to worry about its effect on their hearing. When sixty-six college students were questioned not long ago, forty-five of them said they would be happier if the music in discos were not quite so loud.

"It could still be fun to listen to," said one, "and stop short of causing hearing loss. Maybe a disco owner should have to get a special sort of license, a noise license, that would say just how loud the music could be."

If enough people ask for a law like this, it could be passed.

"The biggest problem in noise control is quiet," said a U.S. Environmental Protection Agency official with a smile. "If we all shouted loudly enough against noise, someone would hear us. The laws we have would be enforced, and new and better ones would be passed."

Of course, the official really did not mean that you should stand on a street corner and shout. What he meant was that people should write letters and join action groups.

For some advice on how to get started, you can write to the U.S. Environmental Protection Agency's Office of Noise Abatement and Control (Washington, D.C. 20460). They have a directory of "Groups That Can Help" (a list of other places to write), and many other booklets.

Your state and city government also try to solve environmental problems, including noise. The agency in charge of air- and water-pollution-control handles noise, too. Late in 1974, for example, the Environmental Protection Administration of New York City made up a special noise package with information. In it were papers describing noise problems and giving advice to citizens on what they can do to control noise. Within a few weeks the agency had to send out fourteen thousand of these packages.

In some smaller cities and towns there may not be an environmental agency. If that is so in your city, write to the Department of Health or the mayor's office.

Look for an antinoise group made up of private citizens in your city or community. If there is one, that is a good place to start. To find such a group, watch for stories in the newspaper about objections and demonstrations against airports and airplanes or against the coming of a highway. Chances are that these are being made by just such a group. There is also a national organization, Citizens Against Noise, with the encouraging abbreviation of C.A.N., which has its headquarters in Chicago. Environmental organizations, such as the Sierra Club and Friends of the Earth, and political groups, including the League of Women Voters of the U.S., are all active in fighting pollution. They have branches in many parts of the country, and there may be one near you.

Perhaps noise has been getting little attention in the place where you live. Your letters and questions can change that. A government agency or national organization may start a community environmental program once they know they can find interested volunteers.

There are many things that young people can do to work toward a quieter environment. They can form an environmental action group in their school. A student committee could meet with teachers and the Board of Education to plan school-assembly programs, showing slides or movies about noise and other forms of pollution. Older students can help to plan programs for kindergartners and first to third graders. Another excellent idea is to arrange field trips where groups of children go out with a tape recorder. They turn it on and record the sound in various parts of town and find out where the noise is loudest and why. Then they can write to the environmental agency, health department, or mayor's office, telling what they have discovered and giving their ideas about what might be done to make things quieter. The Girl Scouts of the U.S.A. has put together a report, *Blueprints for Action,* that has many suggestions like these.

Before taking any action, it is a good idea to learn what is being done about noise control in your city or town right now. What kind of antinoise law has been passed? You can find out by writing to your local environmental agency, health department, or mayor's office. Does the law say anything about automobile horn blowing? About keeping car and truck mufflers in good condition? Banning construction work at night? Setting a time limit for car burglar alarms?

When the New York City noise-control law was being drawn up, two professors at Columbia University asked for a copy of it and passed this around their classes. Each student was asked to put down ideas on how the law could be improved. These ideas were then sent to the mayor and to members of the city council.

Upon hearing of this, a junior-high schoolteacher in a very

noisy part of a New England city got a copy of the local law for his students. To their surprise, they found that the law was a very good one. The problem was that it was not being followed.

"Playing a transistor radio without an earphone on a bus or streetcar is against the law," explained one student, "but practically every time you ride on a bus you hear the sound of music or news."

How many of the noises around you are against the law? They can be reported. And each of us has the power to make the world quieter by small changes in the way we live. Who has not turned the television volume up on a hot day when all the windows are open? Or played a record late at night when the rest of the family was asleep? There are some very simple ways to cut down on noise. If you use a typewriter, get a small foam-rubber pad to put under it. New York City's environmental agency suggests that you wear slippers in the house if you have uncarpeted hardwood or tile floors. Ask a friend to be a dog-sitter when you must go out without your dog. It is not kind to either your dog or your neighbors to leave the animal locked up in the house or car to bark miserably.

"Don't shout from one room to another when a few extra steps would save your lungs and everyone's ears" is some more advice from the New York agency. And across the continent in California, the Sierra Club has put out a bulletin saying simply: "Don't yell at pets or children."

If we followed suggestions like these, life around us would be less noisy. In a truly quiet place the chirp of a cricket carries for half a mile. Where is it that quiet nowadays?

But what if all drivers sounded their horns only to give warning of danger? If houses were built with quieter equipment, and construction workers went off the job at five o'clock each afternoon?

What if trucks and automobiles had good mufflers on their engines and tires that did not screech? What if airplanes were not allowed to land or take off at busy city airports during the night?

Each of these things could happen. The engineers can do the job. Laws have been passed. All of us, tired of noise, are getting behind the efforts at control.

In a quieter world it would be possible to stand at the window and hear not only the rustle of leaves on the trees outside, but also the song of a faraway cricket.

Index

air-conditioning, 19, 62
airplane engine, 56, 57, 61
airplanes, 2, 3, 5, 8, 24, 27, 30, 31, 33, 34, 36, 38, 42, 43, 44, 45, 47, 51, 55, 56, 57, 58, 59, 61, 68, 71, 73, 77
air pollution, 46, 67, 68, 73
airport, 3, 6, 8, 19, 35, 36, 39, 40, 56, 57, 58, 61, 70, 73, 77
Alaska pipeline, 42
animals, 2, 3, 6, 10, 11, 15, 17, 19, 31, 39, 40, 42, 43, 44, 45, 46
annoyance, 2, 22, 33, 36, 38, 47, 61, 67
antinoise demonstrations, 73
apes, 6
appliances, 62, 63, 65
Arctic squirrel, 40
artillery fire, 11
automobile engine, 63, 77
automobiles, 3, 10, 22, 31, 33, 36, 43, 44, 47, 48, 49, 51, 63, 66, 70, 74, 76, 77

babies, 20
balance of nature, 40
bald eagle, 42
"barking-dog laws," 47, 48, 70
bats, 17
bears, 6, 40, 42
birds, 2, 6, 10, 14, 39, 40, 42, 43, 48, 55
blackbird, 39
blasting, 6, 44, 67
Board of Education, Elizabeth, N.J., 72
bobcat, 39, 40
Bogard, Dr. Howard M., 36
bomb, 11
Boulder, Colorado, 6, 69, 70
builders, building, *see* construction
building code, 20
burglar alarm, 74
bus, 26, 30, 66, 76

Caesar, Julius, 3
Canyon de Chelly, Arizona, 59
Cape Kennedy Regional Airport, 39, 40

cap pistol, 11, 26, 31
car. *see* automobiles
castles, 20
cats, 15
chainsaw, 26
chalk, 12, 14
children, 32, 38, 48, 51, 71, 74
Citizens Against Noise, 73
citizens' antinoise groups, 69, 71, 72, 73
city noise, 1, 2, 3, 26, 47, 49, 54, 57, 62, 67, 69
classical music, 25
cochlea, 12, *13,* 26
Colorado University, 69, 70
complaints, 7, 22, 23, 38, 47, 58, 67, 70, 71, 72, 76
concentration, 33, 34
Concorde, 58, 61
confessional, 16
Constitution of the U.S., 61
construction, 2, 5, 6, 7, 19, 30, 44, 47, 48, 66, 67, 68, 74, 76
conversation, 2, 10, 11, 20, 33, 47, 49, 54
Corti, organ of, 12, *13,* 26, 27, 30
cricket, 76, 77
cycles per second (cps), 14

dBA, 15
deafness, 2, 24, 25, 26, 27, 29, 30, 31, 49, 51, 52, 67, 72
  temporary, 27, 31
decibels, 8, 9, 10, 15, 26, 29, 32, 33, 38, 42, 44, 49, 50, 51, 52, 54, 56, 63, 65, 72
digestive system, 9, 34
Dionysius, Ear of, 16
disco, 10, 25, 72
dishwasher, 63
dogs, 15, 19, 47, 48, 76
dolphin, bottlenose, 15
dreams, 36
drill, 2, 5, 26, 47, 48

ear, 2, 5, 9, 11, 12, *13,* 15, 16, 25, 26, 30, 31, 32
  muscles of, 12, 26, 27, 30
  protectors for, 25, 29, 52
  ringing in, 31
eardrum, 11, 12, *13,* 30, 66
echo, 17
electrical amplification, 10, 24, 38
Elmendorf Air Force Base, 45
environmental action group, 74
environmental agencies, state and city, 5, 22, 65, 70, 73, 74, 76
environmental organizations, 73, 76
Environmental Protection Administration, New York City, 73, 76
explosion, 30, 58

factories, 2, 3, 26, 34, 40, 51, 52, 54, 63
Farr, Dr. Lee E., 62, 63
fear, 2, 34, 42, 43, 44
fights, 2, 20, 35, 44, 47
fire alarm, 6, 34
firecracker, 30, 31
fire engine, 36
Fletcher, Dr. John L., 39, 43
foam rubber, 54, 65, 76
food chain, 40, 46
foxes, 40, 44
Franklin, Benjamin, 50
frequency, 12, 14, 15, 21, 22, 25, 27, 29, 32, 40, 43
Friends of the Earth, 73
frogs, 31, 43

Galton, Sir Francis, 15
garbage disposal unit, 63
Girl Scouts of the U.S.A., 74
Goodfriend, Lewis S., 62, 65
Grand Canyon, 8, 9
ground, 22, 43
Ground Control Center, Cape Kennedy, 67

guitar, 5, 11, 20, 24, 38
guns, 25, 26, 29, 30, 31

health department, city, 71, 73
hearing, 2, 9, 11, 12, 15, 16, 17, 22, 24, 25, 26, 27, 29, 30, 31, 34, 51, 72
  of girls and women versus of boys and men, 30, 49
  loss of, *see* deafness
  nerve, 12, 26, 27, 29, 30
  tests, 25, 29, 31, 51
heart, 9, 10, 33, 34, 35
helicopter, 34, 42, 57, 58
hertz (Hz), 14, 15
Hertz, Heinrich, 14
hibernating animals, 40
high blood pressure, 34, 35
highway, 8, 9, 21, 22, 23, 26, 36, 47, 49, 71, 72, 73
home furnishings, 20, 21, 76
horns, automobile, 6, 33, 48, 68, 74, 76
hospitals, 1, 32, 35
hot rod, 66

Icarus, 55
industry, 2, 3, 5, 29, 45, 50, 51, 52, 54
infrasound, 15
insects, 40, 43
Institute of Medical Sciences, 25

Japan, 57, 71
jungle, 3, 21

Khartoum, 35
kitchen noise, 63
Kitty Hawk, N.C., 55, 56
knights, 1, 20, 70
Knights Hospitalers of St. John of Jerusalem, 1

law enforcement, 6, 26, 54, 57, 66, 70, 76
laws, noise-control, 3, 5, 6, 20, 26, 47, 48, 49, 50, 52, 54, 56, 59, 61, 66, 67, 69, 70, 72, 74, 76, 77
  in Boulder, Colorado, 6, 69
  in California, 49
  in Chicago, Ill., 48
  in Illinois, 5, 6
  in Memphis, Tenn., 6, 48
  in New York, 49
  in New York, N.Y., 5, 48, 74
  in Pennsylvania, 48
  in ancient Rome, 3
  in Sybaris, Italy, 2
  in Texas, 48
  in U.S.A., 50, 56, 59, 61, 66, 67
lawsuit, 44, 45, 52, 72
League of Women Voters of the U.S., 73
learning, 33, 72
Lipscomb, Dr. David, 29
Logan Airport, Boston, 35, 61
logarithms, 9
Los Angeles, Calif., 3, 8, 9
Los Angeles airport, 3, 8

Mabaan, 31, 34, 35
Mach, Ernst, 58
Mach 1 (2, 3), 58, 61
machines, 3, 6, 31, 34, 40, 42, 43, 50, 52, 62, 63, 65
manufacturers, 2, 49, 50, 56, 63, 65
mating call, 15, 43
Memphis, Tenn., 6, 48
mental illness, 35, 36
Middle Ages, 1, 3, 20, 70
military equipment, 65
mink, 44, 45
Model Community Noise Control Ordinance, 70
motorcycle, 5, 25, 26, 29, 31, 62, 66
Moussorgsky, Modest, 25
mufflers, 49, 65, 66, 70, 74, 77

music, 3, 5, 6, 10, 15, 24, 25, 26, 27, 45, 48, 52, 76
musical scales, 14

National Institute of Municipal Law Officers, 70
nervousness, 2, 3, 7, 35, 36, 38, 44, 67
noise:
  definition of, 3, 5, 6
  license for, 72
  pollution, 67, 68, 73
  reaction to, 2, 34, 44, 45
  standards for, 50, 51, 67
Noise Control Act, 50
noise-study laboratory, 25
No tobanaide, 71

Occupational Safety and Health Act, 52
Oklahoma City, Okla., 38
old age, 24, 26, 29, 31, 35, 38, 71
Organ of Corti, *see* Corti, organ of
Osaka Airport, Japan, 57, 71

pain, 10, 31, 32
party noise, 6, 10
peacocks, 6
pitch, *see* frequency
plants, 10, 21, 22, 23, 40, 45, 46, 77
power mower, 26, 31, 63
prehistoric cliff homes, 59

"quietest city," 48
Quiet Week, 71

radio, 5, 26, 38, 48, 52, 76
railroad trains, 45, 66
rat, cotton, 2, 3
reading tests, 33
reindeer, 42
reproduction, 42, 43
Rhodes, island of, 1
rivet, 31, 48, 67, 68
robin, 39, 40
rocket, 56, 66
rock music, 5, 10, 24, 25, 26, 27, 29, 38, 45, 72
Rome, Italy, 3, 5
rooster, 2, 48
Rosen, Dr. Samuel, 31, 35

schools, 3, 10, 12, 19, 20, 29, 33, 34, 35, 36, 54, 57, 69, 71, 72, 74
sea life, 59
747 airplane, 56
shrimp, 44
Sierra Club, 73, 76
siren, 14, 27, 36, 43
skunk, 39, 40
Slawinski, Matthew, 52
sleep, 2, 6, 36, 38, 47, 57, 76
slingshot, 71
snoring, 10
snowmobile, 26, 31, 40
sonar, 44
song, 2, 14, 17, 77
sonic boom, 58, 59, 61
sound:
  speed of, 17, 19, 58, 61
  unwanted, *see* noise
sound-measuring equipment, 8, 15, 50
soundproofing, 20, 21, 22, 52, 54, 56, 63, 65, 66, 67, 72
sound waves, 11, 12, 14, 15, 16, 17, 19, 20, 27, 58
spaceship, 55
space shot, 66
speech, 2, 11, 12, 15, 16, 17, 19, 21, 27, 29
speed limit, 66
speed "thermometer," 66
starling, 39
stereo, 2, 25, 27, 30
stomach ulcers, 34, 35
Strutt, John William, 3
students, 10, 12, 19, 24, 27, 29, 32, 33, 35, 36, 48, 56, 69, 71, 72, 74, 76

suburban noise, 2, 47, 49, 57, 70
subway, 26, 27, 30, 33
Sudan, 31, 34, 35
supersonic transport planes (SST), 38, 58, 59, 61
Sybaris, Italy, 1, 2, 3
Syracuse, Italy, 16

tape recorder, 49, 72, 74
teen-agers, 5, 24, 25
television, 2, 20, 26, 30, 36, 38, 49, 56, 76
temperature, 19
Tennessee, University of, 25, 29
terns, sooty, 42
Texas, University of; Graduate School of the Biomedical Sciences, 62
thunder, 3, 10, 42
tires, 6, 68, 77
torture, 32
traffic, 2, 3, 8, 22, 36, 69, 70, 72
trucks, 2, 22, 26, 36, 47, 49, 51, 63, 66, 74, 76
Tupolev-144, 58
Turkey, 43

ultrasound, 15
U.S. Environmental Protection Agency, 25, 26, 34, 39, 50, 65, 66, 67, 70, 72, 73
  Office of Noise Abatement and Control, 73
U.S. Federal Aviation Administration, 59, 61
U.S. Forestry Service, 26
U.S. Navy cruiser, 44
University of Tennessee, 29

vacuum cleaner, 62, 63
voice, 3, 10, 11, 16, 20, 21, 27, 29, 33, 47, 52, 76

warning signal, 2, 6, 43, 44, 48, 76
washing machine, 62, 63, 65
water pollution, 67, 73
whale, 15, 44
whistle, 15, 67
white noise, 20, 21
wilderness, 10, 40
Williams Company, J. H., 52
Woods Hole Oceanographic Institute, 59
World War II, 30, 44
Wright brothers, 55

zoning, 2
zoo, 6

ABOUT THE AUTHOR

Lucy Kavaler is a noted science writer, author of thirteen books and many magazine articles. Her previous books for young readers include another on the subject of environmental pollution, Dangerous Air; *two on the subject of cold,* Life Battles Cold *and* Cold Against Disease; The Wonders of Fungi; The Wonders of Algae *(an A.L.A. Notable Book); and* The Artificial World Around Us.

Ms. Kavaler lives in New York City.

ABOUT THE ARTIST

Richard Cuffari has been illustrating books for more than ten years. A number of his illustrations have appeared in the design shows of the American Institute of Graphic Arts and in the annual exhibits of the Society of Illustrators. He is currently teaching a book-illustration course at Parson's School of Design in New York City.

A native of New York, Mr. Cuffari studied at Pratt Institute. He lives in Brooklyn with his four children.